This edition published by Yosif Limited Ltd in 2018

Yosif Limited Ltd
27 Lilburn Crescent
Massey, Auckland 0614
New Zealand
www.yosif.co.nz
admin@yosif.co.nz

Printed by CreateSpace, An Amazon.com Company

Solar System for Kids

Our solar system contains just one star which is the Sun with eight planets

MERCURY VENUS EARTH MARS JUPITER SATURN URANUS NEPTUNE PLUTO

1

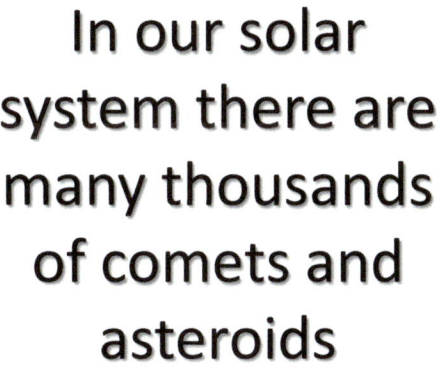

In our solar system there are many thousands of comets and asteroids

Pluto called dwarf planet since it is too small

MERCURY VENUS EARTH MARS JUPITER

SUN SATURN URANUS NEPTUNE PLUTO

STARS COMET ASTEROIDS METEORITE

2

The Sun

The Sun is our galaxy star and the most important object in the Solar System

The Sun gives all the planets in our galaxy heat and light

The Sun

Radius

695,700 Km
432,300 miles

This is Earth

It would take **1.3 million Earths** to fill up the Sun

- Age: 4.6 billion years
- Distance from Sun: 0
- Distance from Earth: 146M km = 91M miles
- Moons: 0
- Mass: 333,000 Earths
- Gravity: 28xEarth
- Surface temperature: 5,500C = 10,000F

The Sun

Missions to the Sun

Ulysses space probe, launched in 1990 and reach the Sun in 1994. Sent info about layer and rays

SOHO space probe, has been studying the Sun since 1995

Mercury

Mercury is the smallest and fastest planet in out solar system. It travels at 48 Km per second!

Mercury name from the Roman god of trade

Mercury

Radius

2,440 Km
1,516 miles

1 Earth = 18x Mercury

- Age: 4.5 billion years
- Distance from Sun: 58M km = 36M miles
- Distance from Earth: 91M km = 57M miles
- Moons: 0
- Mass: 0.055 Earths
- Orbital period: 88 days
- Temperature: -173C (-279F) to 427C (801F)

Mercury

Missions to Mercury

Mariner 10 space probe, It arrived there in 1974 and flew by the planet 3 times. It was able to photograph almost half of Mercury's surface.

Venus

Venus is a very hot planet and it has the longest rotation period of any planet in the Solar System

Venus name from the Roman god of love and beauty

Venus

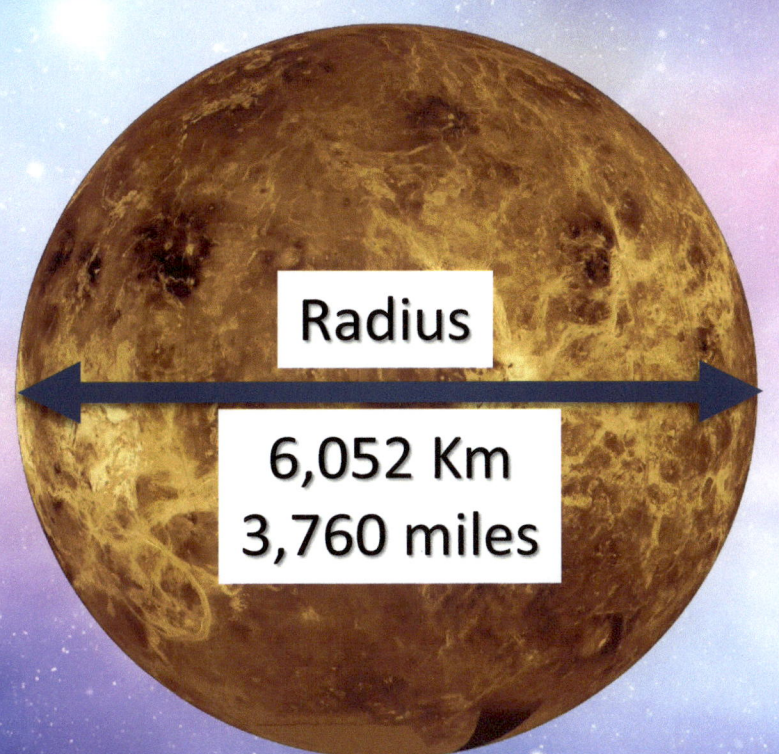

Radius
6,052 Km
3,760 miles

1 Earth = 1x Venus in size

- Age: 4.5 billion years
- Distance from Sun: 108M km = 67M miles
- Distance from Earth: 41M km = 25M miles
- Moons: 0
- Mass: 0.815 Earths
- Orbital period: 224 days
- Temperature: 462C (864F)

Venus

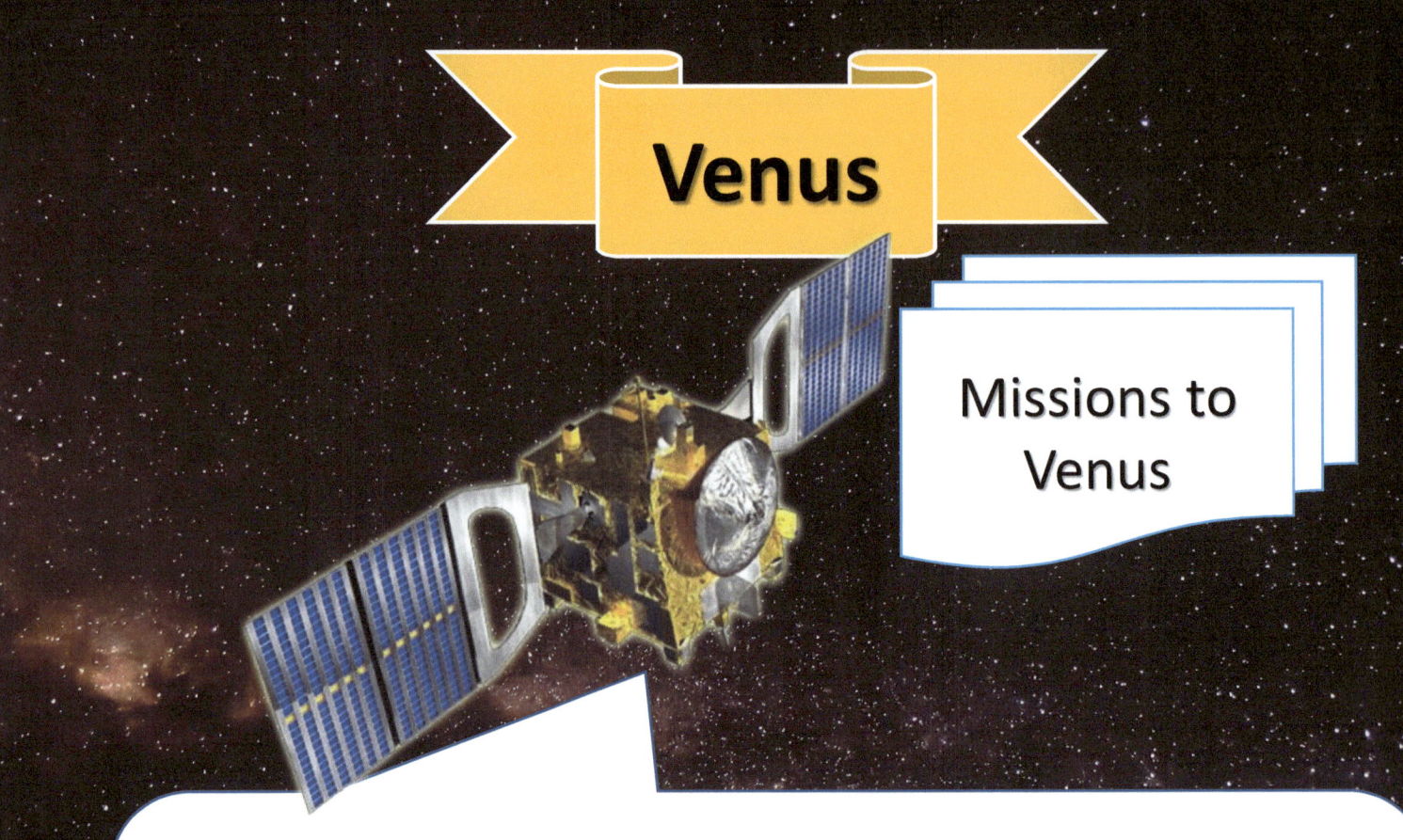

Missions to Venus

Venus Express space probe, scientists would like to send it in the future to study more of Venus' atmosphere and look for microbes.

Earth

The Earth is the only known planet in our solar system with liquid water on its surface

The Earth is out beautiful planet and over three quarters of our planet is covered by water

Earth

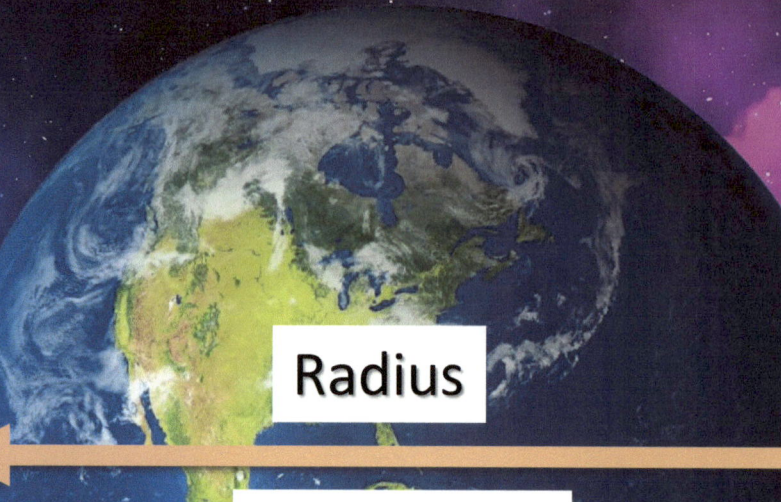

Radius
6,371 Km
3,959 miles

- Age: 4.5 billion years
- Distance from Sun: 150M km = 93M miles
- Distance from Earth: 0
- Moons: 1
- Mass: 0 Earth
- Orbital period: 365 days
- Temperature: 15C (59F)
- Population: 7.5 billion humans

The Moon

The Moon is much smaller than the Earth and it has no atmosphere and no life.

The Moon surface is rocky, dry and dusty and it is in orbit around the Earth

The Moon

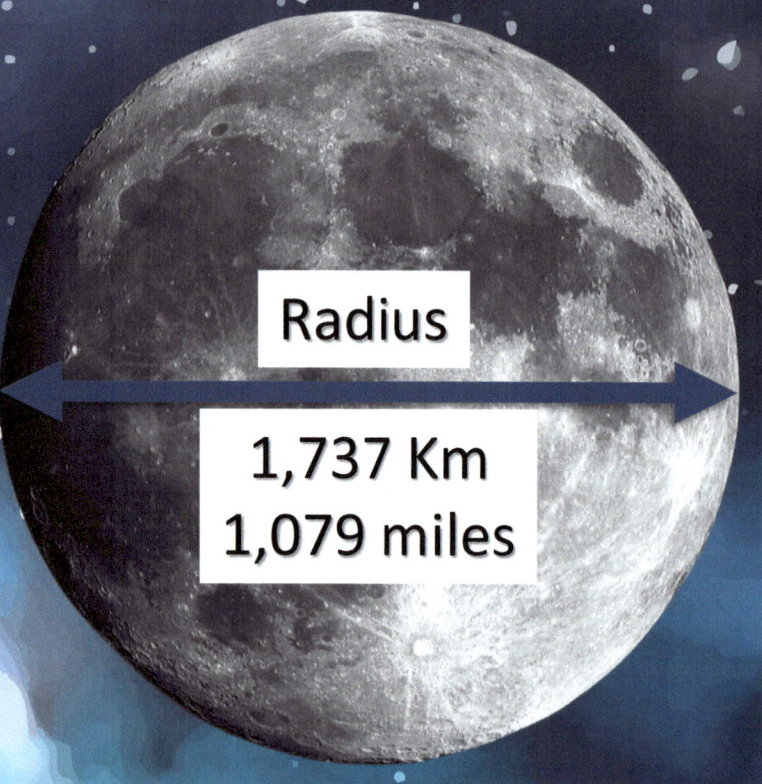

Radius

1,737 Km
1,079 miles

The Moon is less than half the size of the Earth

- Age: 4.5 billion years
- Distance from Sun: 147M km = 91M miles
- Distance from Earth: 400K km = 245K miles
- Moons: 0
- Orbital period: 27 days
- Temperature: -173C (-280F) to 127C (260F)

The Moon

Missions to the Moon

Six Apollo missions have landed on the Moon. First one was Apollo 11 in 20th July 1969 and the last mission was Apollo 17 in 1972. With Apollo 11, the first man on the Moon was Neil Armstrong and the second man was Buzz Aldrin.

Neil Armstrong

Mars

The Mars soil is red because there is iron in the soil and because of that it called the "Red Planet"

A day on Mars is just 41 minutes longer than on Earth days. Mars name from the Roman god of war

Mars

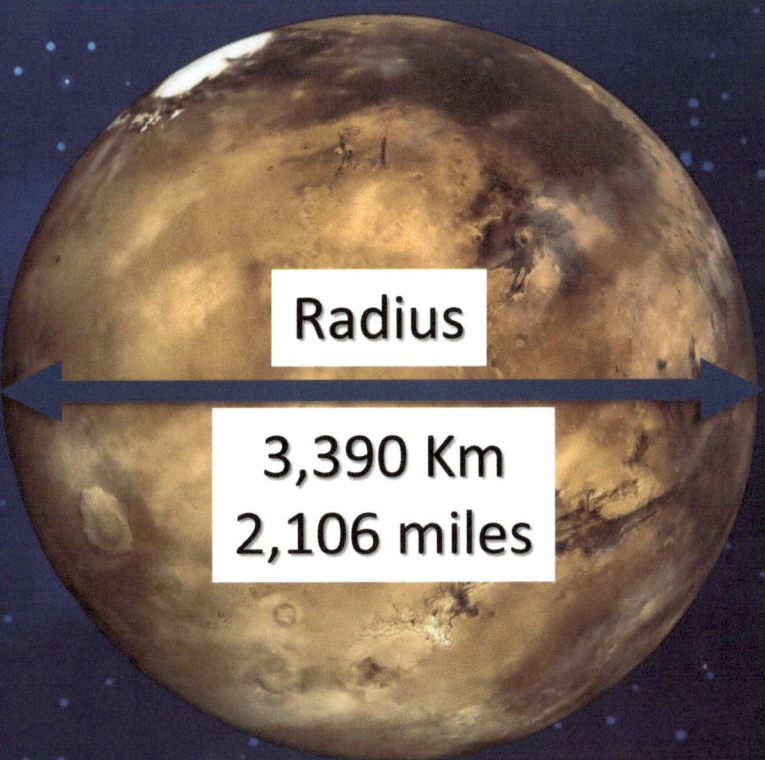

Radius

3,390 Km
2,106 miles

Mars is almost half the size of Earth

- Age: 4.6 billion years
- Distance from Sun: 228M km = 142M miles
- Distance from Earth: 78M km = 48M miles
- Moons: 2
- Mass: 0.1 Earths
- Orbital period: 687 days
- Temperature: from -125C (-195F) to 20C (70F)

Mars

Missions to Mars

Mariner 4 took the first close-up pictures in 1965

Viking 1 landed on Mars in 1979 and took pictures

Robot vehicle to took pictures and tested rocks in 1997

Rover robot from 2004 till now is looking for signs of water

The Asteroid Belt

Asteroids are small rocky objects that orbit the sun. Most asteroids are less than 2 Km across, but some are much bigger

Most asteroids belt between Mars and Jupiter but some of them orbit much closer to Earth

The Asteroid Belt

- Age: 4 million years
- Distance from Sun: 249M km = 155M miles
- Distance from Earth: 100M km = 61M miles
- Moons: 0
- Gravity: 1/100,000 Earths
- Orbital period: 3-6 Earth year
- Temperature: -67C (98F)
- Size: Most of them are relatively small, from the size of boulders to a few thousand feet in diameter.
- There is evidence that some asteroids contain water

Jupiter

Jupiter is the biggest known planet in our solar system. Jupiter is made of hydrogen gas but it has a rocky centre

Jupiter is made of the same material as the Sun and it is the fastest spinning planet.

Jupiter

Radius

69,911 Km
43,441 miles

More than 1,300 Earth could fit inside Jupiter

- Age: 4.5 billion years
- Distance from Sun: 778M km = 484M miles
- Distance from Earth: 628M km = 390M miles
- Moons: 64
- Mass: 318 Earths
- Orbital period: 12 years
- Temperature: -145C (-234F)

Jupiter

Missions to Jupiter

Pioneer 10 in 1973 and took photographs

Galileo in 1989, took pictures of volcanoes, storm and the red spot

Voyager 1 and 2 in 1979, took photographs of atmosphere and its moons

Saturn

Saturn has seven rings around it made of dust, rocks and ice and they are very large and bright.

It spins so quickly that its days around 10 hours. Saturn name from the Roman god of agriculture

Saturn

Radius
58,232 Km
36,184 miles

More than 764 Earth could fit inside Saturn

- Age: 4.5 billion years
- Distance from Sun: 1400M km = 886M miles
- Distance from Earth: 1275M km = 792M miles
- Moons: more than 60
- Mass: 95 Earths
- Orbital period: 29.5 years
- Temperature: -178C (-288F)

Saturn

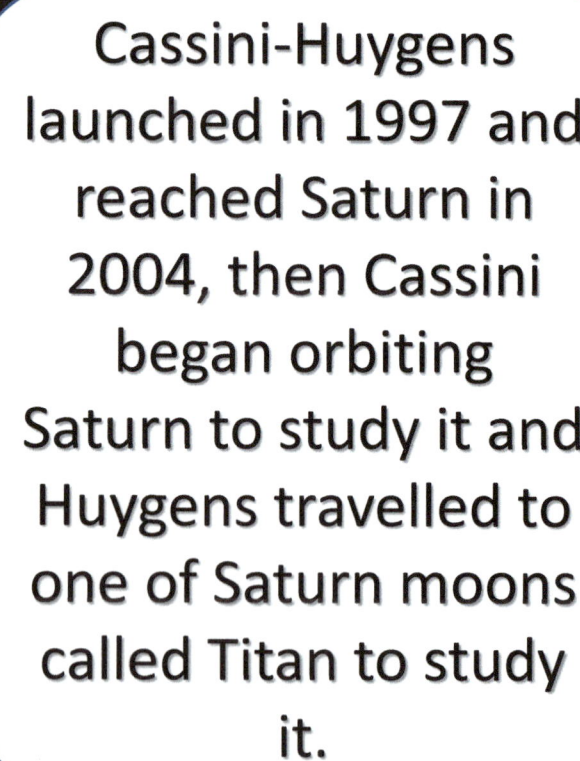

Missions to Saturn

Cassini-Huygens launched in 1997 and reached Saturn in 2004, then Cassini began orbiting Saturn to study it and Huygens travelled to one of Saturn moons called Titan to study it.

Uranus

Uranus has nine rings around it made of dust and rocks. Uranus has storms produce steel and diamond.

It spins so quickly that its days around 17 hours. Uranus name from the Ancient Greek god of the sky

Uranus

Radius

25,559 Km
15,882 miles

- Age: 4.5 billion years
- Distance from Sun: 2900M km = 1800M miles
- Distance from Earth: 2724M km = 1692M miles
- Moons: more than 27
- Mass: 14.5 Earths
- Orbital period: 84 years
- Temperature: -216C (-357F)

More than 63 Earth could fit inside Uranus

Uranus

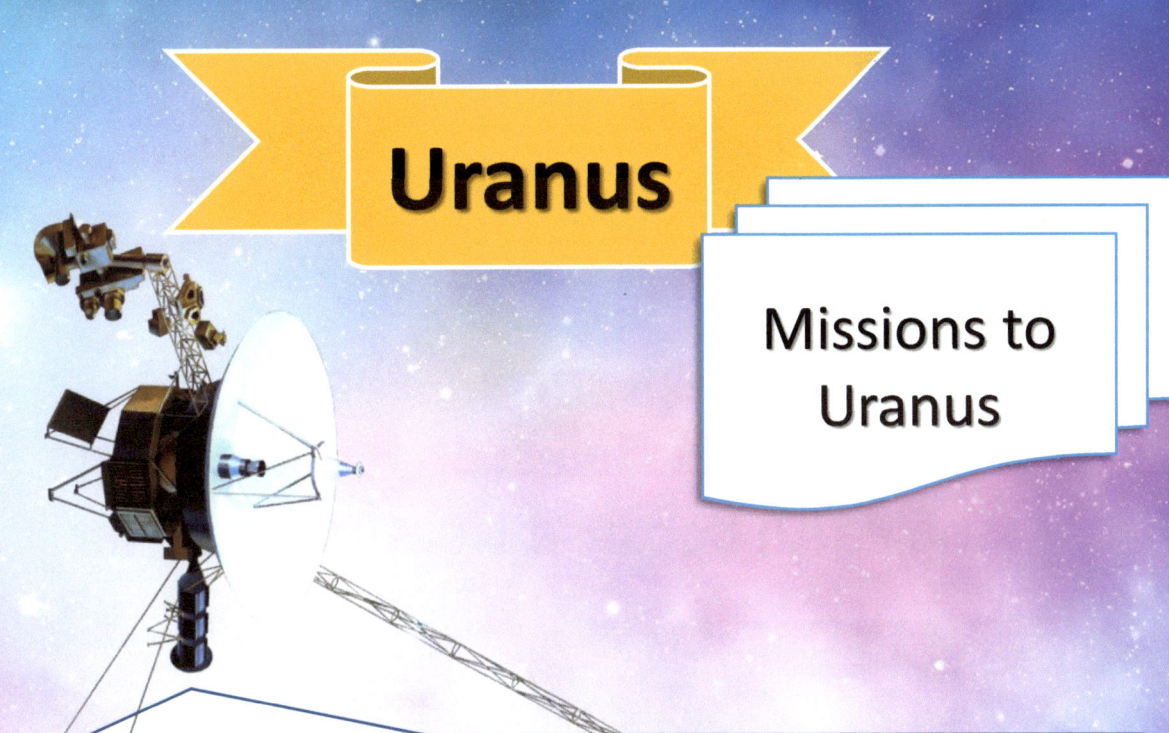

Missions to Uranus

Voyager 2 launched in 1977 and reached Uranus 1986 to study it and now it travelled on towards Neptune and the edge of the solar system

Neptune

To see Neptune we should use Hubble Space Telescope which orbits the earth because Neptune is so far.

It spins so quickly that its days around 16 hours. Neptune name from the Roman god of the sea

Neptune

Radius

24,764 Km
15,388 miles

More than 58 Earth could fit inside Neptune

- Age: 4.5 billion years
- Distance from Sun: 4500M km = 2800M miles
- Distance from Earth: 4351M km = 2704M miles
- Moons: 13
- Mass: 17.1 Earths
- Orbital period: 164.8 years
- Temperature: -214C (-353F)

Neptune

Missions to Neptune

Voyager 2 launched in 1977 and reached Neptune 1989. Voyager 2 measured the temperature on Neptune and the speed of the wind and now it is moving towards the edge of our solar system.

Pluto

Pluto is too small and it is a dwarf planet. Pluto is part of the Kuiper Belt which is a group of objects far out.

Pluto may have a subsurface ocean. Pluto name from the Roman god of the underworld

Pluto

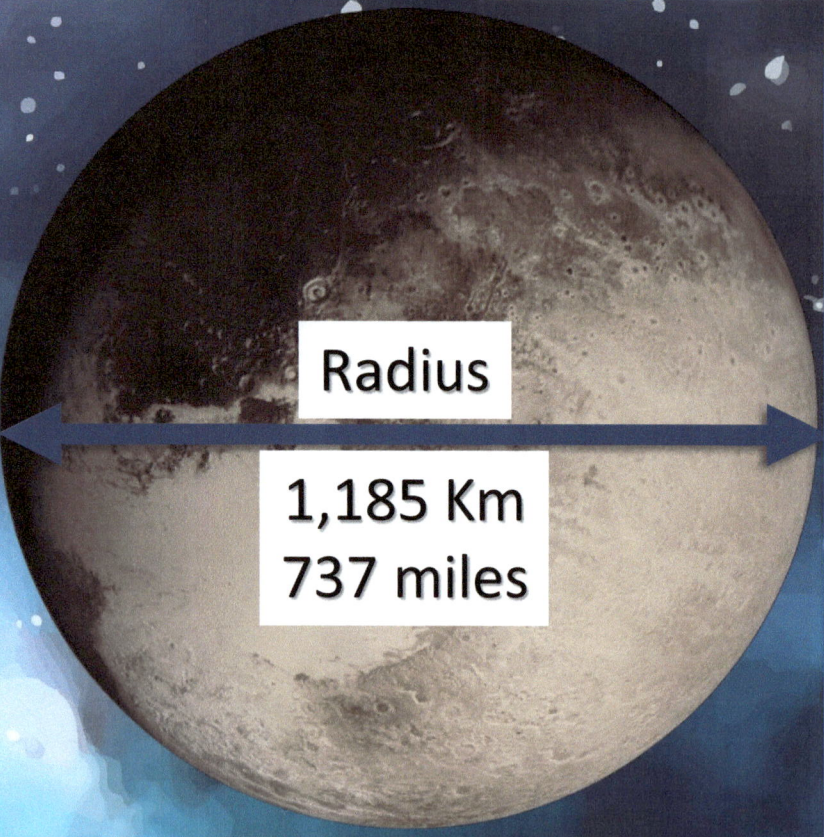

Radius
1,185 Km
737 miles

More than 150 Pluto could fit inside Earth

- Age: 4.5 billion years
- Distance from Sun: 5906M km = 3670M miles
- Distance from Earth: 5756M km = 3577M miles
- Moons: 5
- Orbital period: 248 years
- Temperature: -225C (-375F)
- Pluto is one third water